BRITISH MOTORWAYS – AN INTRODUCTION

MARK CHATTERTON

PRINTED BOOK VERSION

ISBN: 9781910811610

First published in the UK in 2016 by Hadleigh Books
This version published in 2021

This book is also available in the following formats:-
Mobi version for Kindle ISBN: 9781910811580
E-Pub version for Apple I-Pad/Tablet or Kobi
ISBN: 9781910811597
PDF version for personal computers
ISBN: 9781910811603

www.hadleighbooks.co.uk

All text is copyright © Mark Chatterton 2016/2021

Photographs are copyright:
© Mark Chatterton/Wendy Chatterton 2016/2021

All rights reserved.

The rights of Mark Chatterton to be identified as the Author of the Work have been asserted by him in accordance with the Copyright, Designs and Patents Act 1988. A catalogue record for this book is available from the British Library.

All rights reserved. No part of this book may be used or reproduced in any manner whatsoever, nor by way of trade or otherwise be lent, resold, hired out, or otherwise circulated in any form of binding or cover other than that in which it is published, without the publisher's prior written consent.

Disclaimer: All the text within this book is for information purposes only. Neither Hadleigh Books nor the author can be held responsible for any inaccuracies or errors within the text.

Front cover:

The M25 looking north between junctions 28 and 29 near Brentwood in Essex

INTRODUCTION

When I was a child, one of my first "long journeys" in a car, was to go on the newly built M6 motorway between Warrington and Preston, to visit relations on Boxing Day in the 1960's. This became an annual trip for several years and was always an exciting event for a child such as myself. I remember my father never drove faster than 60 mph on the motorway, but it seemed very fast back then, on what by today's standards would be a pretty quiet motorway. Now fifty years later the M6 has become just one of many motorways on the British road network, and I myself have travelled on all of them, either as a driver or as passenger, over the years.

For the past ten years I have travelled all over the country, visiting each one of them and photographing them with a view to writing a book on motorways. This, as the title suggests is an introductory book on Britain's motorways, giving a brief glimpse into some of the most important motorways, as well as some other interesting facts about them. I will of course be writing a larger book at a later date, which will go into greater detail about this interesting civil engineering phenomenon, which is an often taken for granted part of the British landscape. For now, I hope you learn something you didn't know about this vast subject, which affects most of us on a daily basis.

CONTENTS

Introduction

1	What is a Motorway?	1
2	How Motorways are numbered	3
3	The Oldest Motorways	6
4	The Longest Motorways	8
5	The Shortest Motorways	10
6	The Widest Motorways	12
7	The Narrowest Motorways	14
8	The Highest Motorways	15
9	The Busiest Motorways	16

10	The M1	18
11	The M4	20
12	The M5	23
13	The M6	26
14	The M8	29
15	The M25	31
16	The M62	34
17	The A1 (M)	38
18	The A38 (M)	40
19	The A57 (M)	42
20	The A167 (M)	44
21	Former Motorways	46
	List of Motorways	48
	Website Links	52
	Other Books by Mark Chatterton	54
	About the Author	56

BRITISH MOTORWAYS – AN INTRODUCTION

1 - WHAT IS A MOTORWAY?

In 2019 it was estimated that the total length of roadways in Britain was approximately 247,000 miles. Of these there were approximately 32,000 miles of major road, yet only just under 2,300 miles of this total was actually made up of motorway. This seems quite hard to take in considering that you find motorways in all three countries that make up Great Britain. Perhaps this is due to the fact that we assume that there are more motorways than we think, due to the speed it takes us to cover substantial distances when compared with other roads.

What is it that makes a motorway different to all the other types of road that you find in Britain? There are several characteristics which make a motorway different from other roads on the British road network. Here is a definition of a motorway. "A motorway can be defined as a major road which carries motor vehicles in a non-stop environment, free of traffic lights, stopping junctions, roundabouts or driveway access to properties". In practice this should mean that once you enter a motorway, your journey should be a continuous journey with no stopping on the carriageway - apart from an accident, break down or buildup of traffic.

Most, but not all motorways, usually have a hard shoulder on the left hand side of the carriageway for emergency stopping. Motorways should also be free of pedestrians, animals, bicycles, invalid carriages and mopeds.

A typical British motorway scene. This is the M3 in Hampshire looking east from the Fleet services. Note the usual characteristics of a motorway, such as three lanes, a hard shoulder, a crash barrier – in this case made of concrete - and an exit lane to the services in the background on the right.

So, having understood the basics of what a motorway is, let's now look at some interesting facts about our motorways, and how the motorway numbering system works.

2 - HOW MOTORWAYS ARE NUMBERED

A popular myth is that motorways are numbered in the order they were built, but that is not true. The numbers motorways are given is similar to, but not exactly the same as the numbering system for the A roads in Britain that was introduced in the 1920's. This had six different zones working clockwise round a circle beginning at the very top. This is best thought of by thinking of a wheel with spokes running all round it from a centre point. The centre point is London and if you start with a line representing the number 1 going straight up in a northerly direction as far as Edinburgh, you get the start of the A roads section which all begin with the digit 1. Hence the A1 London to Edinburgh trunk road. So any A roads which begin with a number 1 will be to the right of this line E.g. A10, A12, A127.

This zone would be as far as the next zone – the "2 Zone", where the number 2 goes eastwards (south of the river Thames) to Dover. All roads south of this "line" should then begin with a number 2 as far as the next zone. The next number - 3 goes southwards (roughly) to Portsmouth; the number 4 goes westwards to Bristol; the number 5 goes north westwards to the port of Holyhead and the number 6 goes roughly northwards to Carlisle.

All roads in these numbered zones should begin with the number they were in, or at least where they start from. For example, the A38 begins in Devon in Zone 3, but can still be found in Zones 4, 5 and 6, as it runs in a south west to north east direction. For Scotland the numbers start at Edinburgh with a 7 going south west, an 8 going north west and a 9 going north.

Motorways also use a similar system to this, but based on which zone the motorway starts from, would dictate which number it would be given. The M1, being the first long distance motorway, is the first "spoke" in the wheel, going more or less northwards from London. All other motorway zones are then worked out from this in a clockwise direction. So south of the Thames would be number 2 zone, then 3 zone, then 4 zone etc, with Scotland having the 7, 8 and 9 zones.

The main difference with the motorway system to the A/B roads system is that Motorway Zone 4 stops at the M5, so any motorway that starts west of this should begin with a 5. However, there are three exceptions to this – the M48 (previously the old M4 route over the river Severn), the M49, and the A48 (M) which goes into Cardiff from the M4. Their numbers come from the parent motorway – the M4, which runs into South Wales. Plus Zone 5 continues north and west of the M6, as far as the Scottish border.

In Scotland there doesn't seem to be motorway zones as such, apart from the fact that the motorways there do tend to follow the A road they were built to replace. E.g. The M8 follows the route of the A8. There is no M7 as you might expect, but instead you have the M74/A74(M) going south from the Lowlands area, following the route of the A74.

A typical blue motorway sign on the southern section of the M25 within Motorway Zone 2, hence the numbering of the M20. Road numbers, place names and the distance in miles to the next junction are all shown. It is situated well over six feet above the carriageway, so it can be seen at speed from a distance. Note the number junction in white on a black background in the bottom left hand corner. The orange square indicates a diversionary route should the motorway be closed.

3 - THE OLDEST MOTORWAYS

Although Italy and the United States had been building motorways since the 1920's, it wasn't until the 1950's that Britain got its first ever purpose built motorway. It was the rather short 8 ¼ mile long Preston by-pass in Lancashire, which became the first section of the M6. It was built to relieve the traffic jams which were regularly occurring in the town. It was opened on December 5 1958 by Harold Macmillan, the then Prime Minister and had a hedge on the central reservation to cut down glare from oncoming traffic headlights. It had two lanes in either direction, with a wide enough central reservation for third lanes to be added at a later date. This did in fact occur in 1966; a year after the M6 motorway incorporated the original Preston by-pass into its route.

The next motorway to be opened was in fact three different motorways all linked together. This was on November 2 1959 when the M1 was opened from near Watford (Junction 5) to Crick, near Rugby (Junction 18). There were also two spurs at either end to help distribute traffic. These were the M10 at its south end going south to St Albans, and the M45 at its northern end, going towards Coventry.. (See M1 section below for more details).

Note: Although some people argue that the second motorway to be opened was a short section of the M4, known as the Chiswick Flyover in September 1959, it was not originally a motorway, but part of the A4. It wasn't until 1964 that the flyover became part of the M4.

The M45 in Northamptonshire, looking west. Being one of the earliest motorways, it still has many of its original concrete over bridges, designed by Owen Williams, which were built over sixty years ago. It is also one of the quietest motorways in Britain. Note the central reservation crash barrier hidden by the long grass!

4 - THE LONGEST MOTORWAYS

The longest motorway in Great Britain is the M6, which is 232 miles (373 km) in length. It starts at Catthorpe, near Rugby in Warwickshire, where it leaves the M1 at Junction 19 and goes via Birmingham and the North West as far as Gretna, by the Scottish border, where it joins to the A74 (M). It could be argued that it also has the accolade of the longest time taken for a motorway to be constructed. The first section was opened in 1958 as the Preston by-pass but the last section between Carlisle and Gretna, just over the border in Scotland, was opened as recently as 2008. So that makes it fifty years in the making!

If you take in the motorway links both north and south of the M6, you can get an uninterrupted motorway journey of over 460 miles. This would start at Exeter at the south end of the M5, go up to the M6 at Junction 8, and continue up the M6 to the Scottish border. Then you would go straight onto the A74 (M), which then changes into the M74. Then you would veer right just before Glasgow onto the M73, before going onto the M80 and finally take the M9 to its end at Dunblane.

Alternatively, since the second crossing of the Forth was opened, you can now go along the M8 from the M73, onto the M9 and finally go north on the M90 to its end at Perth; a distance of around 510 miles.

The next longest motorway after the M6 is the M1 which runs from North London to the A1 (M) to the north east of Leeds. This motorway is 193.5 miles long (311.4 km). However, since the section of the A1 (M) that it joins up with is now a motorway as far as Newcastle-upon-Tyne, (via the A194(M)), the total length of continuous motorway here is now just under 300 miles (482 km)

The start of the M6 in Leicestershire as it leaves junction 19 of the M1 at Catthorpe, on its 232 mile journey to the Scottish border.

5 - THE SHORTEST MOTORWAYS

There is a debate about which is the shortest piece of motorway in Britain. Some say it is the A635 (M) in Manchester. This is the eastern end of the Mancunian Way which is numbered as the A57 (M) on road signs and maps. The A635 (M) isn't actually signposted as such, but was designated a motorway by the Department for Transport in 1995. It is just 0.3 miles in length.

Others argue for the M898 in Renfrewshire, Scotland, as that is a true motorway in numbering, as opposed to the A635(M) which is an A road converted into a motorway. Plus the M898 is a complete motorway at half a mile in length, whereas the A635 (M) is part of another motorway.

Then there is a third contender. With the completion of the missing link of the M8 in Scotland, there is a new contender for the shortest motorway in Great Britain. This can be found at the Baillieston Interchange where the M8 (junction 8) and the M73 (junction 2) motorways meet at a roundabout.

BRITISH MOTORWAYS – AN INTRODUCTION

It is called the A8 (M), and is a link road of motorway standard which is 283m in length. It links the A8 (Edinburgh Road) with the two motorways.

This is a debate which could run and run until someone at the Department of Transport gives an official view on this. For the time being, the author of this book chooses to go with the M898, especially as this is a proper motorway with an "M" prefix as opposed to the M in brackets..

The M898 looking south, shortly before it joins up with the M8 to the west of Glasgow in Scotland. Not only is it the shortest "M" prefixed motorway, but it is also the highest numbered "M" motorway.

6 - THE WIDEST MOTORWAYS

Motorways are usually three or four carriageways wide, though there are some which are just two or even one carriageway's width. There are several which have more than four lanes in parts. The M25 has five lanes between junctions 12 and 14 in Surrey, and six lanes on both carriageways between junctions 14 and 15, between the A3113 (Heathrow Airport Terminal 5) and the M4 interchange. Needless to say this is one of the busiest stretches of motorway in Britain. (See below)

In Glasgow there are two sections of the M8 which are also very wide. As the M8 crosses the River Clyde on the Kingston Bridge, there are five lanes on each carriageway for a short section. Then further south, between junctions 21 and 22 for about a mile there is an even wider section where the M8 converges with the end of the M74 and the start of the M77. Add to this several local roads all converging here, making the motorway and other roads sixteen lanes wide.

Further south in England to the north west of Manchester is the Worsley Interchange. Here the M60 and the M61 motorways meet. The interchange runs for two and half miles between junctions 14 and 15 of the M60, and between junctions 1,2, and 3 of the M61. Add to this some local roads and it becomes a massive eighteen lanes wide at Linnyshaw Moss. Again, like the Glasgow motorways this total includes both motorway lanes and non-motorway roads.

Once again, this a very much an ongoing debate with some arguing that non-motorway roads running next to a motorway don't count as lanes of a motorway.

The M8 and the M77 converging to the south west of Glasgow city centre, on one of the widest sections of motorway in Britain.

7 - THE NARROWEST MOTORWAYS

Until 2020, this was the A601 (M) in Lancashire, which ran south as a spur from junction 35 of the M6 to the B6254 at Over Kellet. It has single lanes in each direction and measures 24 feet or 7.3 metres in width. This was also the only motorway to end at a T-junction onto a B road. In 2020 it was downgraded to become part of the B6254.

A view looking along the A601(M) motorway (as it then was) in Lancashire shortly before it ends at the junction with the B6254.

Since 2020 the narrowest motorway in Great Britain is probably the three lane motorway spur linking the M61 and M65 motorways at Walton Summit in Lancashire.

8 - THE HIGHEST MOTORWAYS

The highest point on a motorway in Britain is on the M62 near Junction 22, where the carriageway reaches a height of 1,221 feet (372m) above sea level. There is a sign proclaiming this at the spot, which is quite desolate to put it mildly and is known locally as "Windy Hill".

The next highest spot on a motorway in Britain is on the M74 in Scotland, south of Glasgow, between junctions 12 and 13 and is approximately 1,060 feet above sea level. Not far behind is Shap Summit at 1,050 feet (320m) above sea level in Cumbria on the M6, between junctions 39 and 40.

The M62 looking west under Scammonden Bridge, between junctions 22 and 23 of the Motorway, not far from the highest point on any motorway in Britain.

9 - THE BUSIEST MOTORWAYS

Whenever we are stuck on a motorway in traffic, it is easy for us to think that we are on the busiest motorway in Britain! The problem is that figures from official sources give out the total number of vehicles in a day as an average, rather than by the hour, so busy times around the two rush hours are averaged out by the quieter times in between. According to the government traffic counts, there are several contenders for the title of the busiest motorway.

The M62 in West Yorkshire is one of the busiest motorways in the north of England, with an average daily traffic flow of 144,000 vehicles recorded around the Bradford/Leeds area in 2008. Move forward to 2014 and the Annual Average Daily Flow (AADF) sees even more cars than this on some of our motorways.

The M60 to the west of Manchester, between junctions 12 and 13 saw a peak of 195,000 vehicles. Further south on the M1 between junctions 7 and 8 there were 197,000 vehicles in an average day.

Yet the motorway with the highest figures of all is the M25, as would be expected. This was between junctions 14 to 15 near to Heathrow Airport, where there were a massive 262,000 vehicles per day recorded in 2014.

Tailbacks on the M62 heading east near Bradford; one of the busiest stretches of motorway in the north of England.

10 - THE M1 – THE GATEWAY TO THE NORTH

The M1 runs from its junction with the North Circular Road (A604) in London to the east of Leeds, where it joins up with the A1 (M), 193 miles (311km) later. It was the second motorway to be built in Britain, soon after a short section of the M6 had been opened. It also had two spur motorways built with it - the M10 at the south end and the M45 at the north end. It was Britain's first purpose built long distance motorway and opened in 1959, clocking in at around sixty miles in length. It was quite revolutionary at the time, as it did not have a speed limit and was treated as a race track by some motorists and motorcyclists.

Plans to build a motorway between London and Birmingham had been hatched back in the 1920's, but it wasn't until the 1950's that it finally got the go ahead. In reality the first section of the M1 motorway went from Watford in Hertfordshire to Crick, near Rugby in Warwickshire. It had two spurs; the M10 going to near St Albans and the M45 going to the south of Coventry. These were designed to take away some of the traffic from each end of the M1 and distribute it to other areas, away from the West Midlands and London.

The M1 north of Rugby was built in various sections throughout the 1960's and was connected to Leeds by 1968. The M1's end in Leeds stayed in place until 1999, and was part of the Leeds South Eastern Motorway. This section north of Junction 43 has now been renamed the M621, whilst the M1 now continues to the east of Leeds to join up with the A1 (M) at Hook Moor, forty-eight junctions later after starting in London. On its journey north it passes by the major towns of Luton, Milton Keynes, Northampton, Leicester, Derby, Nottingham, Sheffield, Wakefield and Leeds. It is connected to the following motorways: - the M6, M18, M25, M45, M62, M621 and the A1 (M).

The M1 near Leicester, looking towards the southbound carriageway from the Leicester Forest East services.

11 - THE M4 – THE ROAD TO THE WEST

The M4 motorway runs from West London to about ten miles north of Swansea in West Wales, where it joins up with the A48. It is just less than 200 miles (320 km) in length and was built in sections between 1959 and 1993. In fact the first section of the M4 to be built, the Chiswick Flyover in 1959 was not originally the M4, but rather an elevated section of the A4. When the section containing junctions 1 to 5 was opened in 1965, the Chiswick Flyover then became part of the M4.

The M4 was originally going to terminate just west of Newport in South Wales at Tredegar Park, but with the introduction of the Welsh Office, the M4 was allowed to go much further into Wales. In fact it eventually went to the western side of Swansea, terminating by Pont Abraham services in Carmarthenshire at junction 49. Here the motorway connects to the A48 and the A483. There is also a spur going from the into Cardiff, the A48 (M).

The Second Severn Crossing pictured above takes the M4 from England into Wales over the estuary of the River Severn. It was opened in 1996, thirty years after the original Severn Bridge. It has since been renamed the "Prince of Wales Bridge".

The M4 was also re-routed at the same time as the above bridge was opened. The new route was more direct and included the new bridge. The previous route of the M4 further north which included the original Severn Bridge Crossing was renumbered as the M48.

In 2018 the road toll payable by vehicles going from England into Wales via both Severn bridges was abolished.

The M4 in England was built in sections between the years 1961 and 1971. The link between England and Wales was joined in 1966 when the original Severn Bridge was opened in by Queen Elizabeth. The Welsh part of the M4 was finally completed in 1993 with the opening of the Britton Ferry Bridge. It was the first motorway in Britain to have a tunnel. This is the Brynglas Tunnel in Newport which was opened in 1967 as two twin tunnels. However, as the M4 has to change from three to two lanes either side of the tunnel, it has become a bottleneck on a daily basis. This has led to plans for a new motorway to be built to the south of Newport, provisionally called "the M4 Relief Road".

As the M4 carries a lot of traffic into Central London from Heathrow Airport, it has its own spur motorway, leading from Junction 4 down into the airport. One bone of contention was the 3 ½ mile long M4 Bus Lane, between Heathrow and the elevated section, which existed on the outer lane of the M4 going into London between the years 1999 and 2010, and in 2012 during the Olympics. It was closed permanently after the Olympics.

On its route west, the M4 passes the towns of Slough, Reading, Swindon, Bristol, Newport, Cardiff and Swansea. It is connected to the following motorways:- The M5, M25, M32, M48, A308 (M), A329 (M), A404 (M),

12 - THE M5 – THE ROAD TO THE SOUTH WEST

The M5 runs south west from Birmingham in the West Midlands to Exeter in Devon. It was built to help relieve the amount of traffic going between the West Midlands and the North of England, and the West Country and South Wales on the busy A38. The first section between Worcester and the M50 was opened in July 1962, with the M50 having been already constructed in the years 1958 to 1962. A section further south, by-passing Filton in north Bristol was also opened in 1962. The section from Worcester north to the M6 in Birmingham was built between 1967 and 1970. The sections south from the M50 through Gloucestershire and Somerset to Devon were built in a ten year period between 1967 and 1977.

Before the M5 was opened all the way from Birmingham to Exeter, one major bottleneck on the journey was the city of Bristol. Until the Avonmouth Bridge was opened in May 1974, traffic had to go either through or around Bristol on this route. Even now there can be hold ups on the bridge as the gradient is quite steep to allow tall ships to pass underneath it. Slow lorries can cause the traffic to build up behind them in the rush hour and during summer weekends.

A typically busy M5 on a summer's evening looking north from Junction 9 for Tewksbury.

One major feature of the M5 is the Almondsbury Interchange to the north of Bristol where the M4 and M5 meet. It is a four level stack interchange (or butterfly junction) with free flowing traffic going in all four directions. When it was first built it was the most complicated junction on the motorway network, but has since become a bottleneck and has become a managed motorway to help relieve congestion.

BRITISH MOTORWAYS – AN INTRODUCTION

The M5 can get very busy during the summer months with holiday traffic going to and from the South West, especially on a Friday and Saturday, plus Bank Holidays Mondays. One point of interest for motorists who might be stuck in a traffic jam south of Bristol is the Willow Man, which stands next to the M5 near Junction 23. It is a 40 feet high sculpture which is made out of willow withies.

In Birmingham to the north of junction 3 the motorway is mainly elevated to go over the two sections of the Birmingham Canal. When it joins up with the M6 south of Walsall this elevated section continues for several more miles, making it one of the longest sections of elevated motorway in Britain.

Starting in north west Birmingham, the M5 passes the towns of Bromsgrove, Worcester, Cheltenham, Gloucester, Bristol, Weston-Super-Mare and Taunton on its route. The M5 is connected to the following motorways:- the M4, M6, M42, M49.

13 - THE M6 – THE ROAD TO SCOTLAND

Britain's longest motorway, the M6, starts at junction 19 of the M1 near Rugby and goes through the West Midlands and the North West of England to the Scottish border at Gretna, where the motorway continues as the A74 (M). The first part of the M6 to be built was the Preston-by-pass, which was the first section of motorway ever to be opened in Britain.

This was in December 1958 and since then various sections have been added mainly in the years 1960 to 1972. By then the motorway stretched from Rugby to Carlisle, but there was one more piece of the M6 to be added. This final section was the "Cumberland Gap" between Carlisle and the Scottish border, where the M6 becomes the A74 (M). This finally happened in 2008, and was officially opened fifty years to the day after the first section of the M6 was opened.

BRITISH MOTORWAYS – AN INTRODUCTION

Due to its length, the M6 passes through all sorts of terrain in both urban and country areas. The section between junctions 36 and 40 that passes through Cumbria is one of the most picturesque parts of the motorway network, taking you through the Lune Gorge and over Shap Fell. Also worth looking out for is the futuristic looking Lancaster Services, which has a UFO type tower in it, called the Pennine Tower. It used to house a restaurant and observation deck, but has been closed to the public since 1989 due to health & safety regulations.

Further south near Warrington, the M6 crosses the River Mersey and the Manchester Ship Canal on the iconic Thelwall Viaduct. It is actually two separate bridges, the first which opened in 1963, and the second which opened in 1995. It carries around 160,000 vehicles per day.

At junction 6 where the M6 meets with the A38 (M) Aston Expressway, joining the motorway to Central Birmingham is the famous "Spaghetti Junction". It's real name is the Gravelly Hill Interchange and was opened in 1972. Apart from the A38 (M), it is also accessed by the A38 (Tyburn Road) and the A5127 (Lichfield Road). It has five different levels and goes over two rivers, three canals and two railway lines.

The M6 going north passes near or through the major towns of Coventry, Birmingham, Walsall, Wolverhampton, Stafford, Warrington, Wigan, Preston, Lancaster and Carlisle. It is connected to the following motorways:- the M1, M5, M6 Toll, M42, M54, M55, M56, M58, M61, M62, M65, M69, A38 (M), A74 (M), A 601 (M).

The M6 looking south from the Charnock Richard services, south of Preston in Lancashire. This section was one of the earliest sections of the M6 motorway to be opened in 1963.

14 - THE M8 – SCOTLAND'S MOST IMPORTANT MOTORWAY

The M8 is Scotland's main motorway and runs mainly between Glasgow and Edinburgh. The motorway starts to the west of Edinburgh at its junction with the Edinburgh City bypass and goes roughly in a westerly direction to Glasgow. It continues through much of Glasgow, going north of the city centre, before turning south over the River Clyde and then going west again, past Glasgow Airport and finishing at junction 31 at Bishopton in Renfrewshire, where the A8 takes over.

It is the busiest motorway in Scotland and with the conversion of the six mile gap on the A8 to the M8 between Baillieston and Newhouse in 2017, is now sixty six miles long. The first section to be opened was the Harthill bypass in 1965, with the section through Glasgow constructed between 1968 and 1972. Other sections were added piecemeal until the motorway that exists today was finally completed in 2017.

The section through Glasgow is notorious for its traffic jams, mainly because of the amount of local traffic that uses it, but also because three motorways converge on each other just south of the Kingston Bridge.

Also worth noting are that there are several right hand exits off the fast lane in Glasgow, which is an unusual feature for motorways in Britain. It has thirty one junctions, a short tunnel at Charing Cross in Glasgow and is on a concrete viaduct for a large section through Glasgow.

It passes nearby or through the following towns:- Livingston, Coatbridge, Glasgow and Paisley. It is connected to the M9, M73, M74, M77, M80 and M898 motorways.

The M8 in Lanarkshire looking east from the Shotts services.

15 - THE M25 – THE ROAD TO HELL?

The M25 is possibly Britain's most notorious motorway, mainly because of the regular number of traffic jams that occur on its 117 mile (188 km) length. The song, *The Road to Hell* by Chris Rea is reportedly written about the M25, plus jokes about it being the biggest car park in Europe are well known. For a time it was the longest orbital motorway in Europe, but the A10 Berliner Ring Autobahn in Germany has overtaken the M25 in length by just five miles.

It reached its thirtieth birthday in 2016, yet it is not a complete circle around London as many people believe. In fact the motorway starts and ends on either side of the Dartford Crossing between Kent and Essex. The road that carries the Dartford Bridge and Tunnels is actually the A282, a rather boring number for such an important piece of road. Just to confuse things, Junctions 1A and 1B are on the A282, whilst the first junction on the M25 is Junction 2 (the A2). At the other end in Essex, the final junction on the M25 is Junction 30 for the A13, yet there is a Junction 31 on the A282 for the A1306. just a mile beyond the end of the M25.

The idea of a ring road around London was first thought of as long ago as the 1930's, which later evolved into a series of ring roads or "Ringways" in and around London in the 1960's. It wasn't until the 1970's that construction work for the M25 began, with the first section opening in 1975, to the north of London connecting Potters Bar (junction 24) with South Mimms (Junction 23). Incidentally, Sir Bob Geldof was one of the many workmen who helped build this section. Further sections were then opened between 1976 and 1986 in various stages all around outer London, with the final stage being Junction 19, Micklefield to Junction 23, South Mimms - where it all began.

When it was finally opened on October 29 1986 by the then Prime Minister, Margaret Thatcher, the M25 had become Britain's most expensive motorway, clocking in at £909 million, which works out at approximately £7.5 million per mile of construction. Since then, at least the same amount has been spent on it, upgrading it, mainly with extra lanes to cope with the amount of traffic that uses it, and turning it into a smart motorway on many sections.

BRITISH MOTORWAYS – AN INTRODUCTION

A big fear soon after it was opened was that some people would use it as a race track to see how long it would take to do a complete circuit of the motorway. This did in fact happen and helped bring in the speed enforcement cameras, with rumours of a lap of the motorway clocking in at below one hour for the 117 miles. The M25 was also well used at weekends in the 1980's and 90's by partygoers on their way to illegal raves somewhere in the Home Counties. The duo, Orbital are said to have taken their name from the M25's other name – "The London Orbital Motorway".

On its route, the M25 passes near to the following towns; Dartford, Sevenoaks, Reigate, Epsom, Woking, Staines, Uxbridge, Watford, St Albans, Enfield, Romford and Brentwood. It is connected to the following motorways: the M1, M3, M4, M11, M20, M23, M26 and M40 and A1 (M).

The M25 between junctions 29 and 30 in Essex looking north.

16 - THE M62 – THE CROSS BRITAIN MOTORWAY

The M62 runs from Liverpool in the west, passing near to Manchester, Bradford and Leeds, before terminating just short of Hull in the east. Until 1998 it was a continuous motorway, 107 miles (172 km) in length, but with the completion of the M60 Orbital Motorway around Manchester in 2000, it was effectively cut in two, as the section between junctions 12 (Eccles) and 18 (Simister) became part of the M60. However on the road signs of this section, the original junction numbers have been kept the same for both motorways to avoid confusion.

It was one of the hardest motorways to construct, as the middle section went right over the Pennine Hills, between Manchester and Leeds and there were delays due to the difficulty of the terrain, coupled with bad weather, taking several years to complete. The current motorway was finally finished in 1976, six years after construction first began. Strangely though, the stretch over the Pennine Hills had been officially opened by the HM the Queen on October 14 1971.

It is both Britain's highest and lowest motorway. It reaches a height of 1,221 feet (372 metres) near Junction 22 on Saddleworth Moor, on the western side of the Pennines. It then reaches a height of just two metres above sea level near Goole at its eastern end.

Back in the Pennines, on the Yorkshire side of the M62, between Junctions 22 and 23, the motorway divides in two as it goes around Stott Hall Farm. A popular urban myth is that the farmer refused to move out when the motorway was being built, so they had to build the motorway around the farm. In actual fact the carriage ways were split due to the ground having weak peat bogs and to go in a straight line through the farm could have resulted in the carriageway sinking into the soft ground.

The M62 also saw the first and only terrorist attack on a British motorway, when a coach carrying military personnel and their families was destroyed by a bomb believed to have been planted by members of the Provisional IRA on February 4 1974. This happened between Junction 26 (Chain Bar) and Junction 27 (Gildersome). Twelve passengers were killed and thirty eight others injured. A memorial to the victims is to be found at the Hartshead Moor services between Junctions 25 and 26. (See picture overleaf).

In March 2018 almost three and a half thousand vehicles were trapped overnight on the eastbound carriageway of the motorway between junctions 20 and 24. This was due to a car fire coupled with extreme weather conditions.

One other thing is that the motorway actually starts at Junction 4 (Queens Drive), on the eastern edge of Liverpool and not Junction 1 in the city centre. This is because it was intended to build the motorway into central Liverpool, terminating near Lime Street Station, just short of the city centre, but in the end this never happened.

The M62 passes near the towns of Warrington, Manchester, Rochdale, Oldham, Halifax, Huddersfield, Bradford, Leeds, Wakefield, Pontefract and Goole. It is connected to the following motorways: - M1, M6, M18, M57, M60, M61, M66, M602, M606, M621, A1 (M), A627 (M).

The memorial to the victims of the terrorist bomb of February 1974, which occurred on the M62 near the Hartshead Moor services, West Yorkshire.

BRITISH MOTORWAYS – AN INTRODUCTION

Apart from motorways with an M prefix, there is another group of motorways which are to be found in Britain which begin with an "A" prefix with an "M" suffix after the road number. E.g. A1 (M) or A66 (M). These are roads which were originally built as A roads, but which have since been upgraded to a motorway, with any private drives or roundabouts removed. In this section we will look at several notable examples of these A road motorways.

17 - THE A1 (M) - THE GREAT NORTH ROAD

The A1 (M) is part of the Great North Road, the A1, which has been upgraded to motorway status. It actually refers to four separate sections of motorway which are situated on the A1 road, between London and Edinburgh.

Starting at its southern end, the first section of A1 (M) motorway is between the M25 at South Mimms, (junction 1) and goes as far as Junction 10. This section wasn't all built at the same time, but at various stages between 1962 and 1986. The second section of motorway on the A1 is between Alconbury and Peterborough (junctions 14 to 17) and was opened in 1998, with two four lane carriageways for much of its 13 miles. The third section, the Doncaster by-pass, between Junctions 34 and 38, opened in 1961. It is one of the earliest sections of motorway in Britain and is two lanes in each direction. The fourth and final section (built between 1965 and 2017) is the Darrington to Gateshead section, (Junctions 40 to 65), which includes the junction where the M1 joins up with the A1 (M) and the well-known Scotch Corner Interchange. In total, there are approximately 150 miles (241km) of motorway that make up these four stretches.

The various parts of the A1 (M) on its route north passes the following towns: - Hatfield, Welwyn Garden City, Stevenage, Peterborough, Doncaster, Leeds, Darlington, Durham and Washington. It connects with the following motorways:- M1, M18, M25, M62, A66 (M), A194 (M).

The Alconbury to Peterborough section of the A1 (M) looking north, soon after the junction with the A14. Note the four lanes of this motorway, which was effectively rebuilt from the previous two lanes each way in 1998.

18 - THE A38 (M) - THE ASTON EXPRESSWAY

The two mile section of motorway which runs from Birmingham City centre to the M6, the A38 (M), is most famous for its junction with the M6, which is more well known as "Spaghetti Junction". The road itself is called "The Aston Expressway" as it goes mainly through the Birmingham suburb of Aston. It is unusual for a motorway, as it does not have a central reservation, though it does have a "buffer lane" between the two flows of traffic.

This is usually one of the middle of the seven lanes that are on this motorway. This is because it uses the system of "tidal flow" whereby up to four lanes are used in the morning rush hour going into Birmingham, with just two coming out. Then in the evening rush hour this is reversed. In off peak periods there are usually three lanes in each direction,

The A38 (M) motorway was opened in 1972 and involved the demolishing of many old houses to make way for the motorway. As it is an Urban Motorway used by a high number of vehicles each day there are frequent traffic jams. This has led to locals calling it "The Aston Distress Way"!

To help with safety and traffic flow there is a 50 mph maximum speed limit for most of its route, though this reduces to 30 mph near the Birmingham end on the southbound carriageway. Unusually at the time of writing there are no speed cameras on the motorway.

The start of the A38 (M) Aston Expressway in Birmingham, looking north. Note the 50 mph speed limit on the blue road sign, which is unusual for motorways in Britain.

19 - THE A57 (M) - THE MANCUNIAN WAY

The A57 (M) is better known as "The Mancunian Way" and it passes along the southern edge of Manchester city centre on a two mile elevated motorway, going from west to east. It was opened in May 1967 by the Prime Minister at the time, Harold Wilson, as an A road and upgraded to a motorway in the 1970's.

It is unusual in that there is a slip road that goes nowhere on the north side of the road, which is hidden by a large advertising board. It is actually the end of a road that was supposed to go into the centre of Manchester and ends twenty feet above the ground!

Further along the A57 (M) some people have argued that it changes into the A635 (M) at its eastern end, as the A57 continues off the motorway at its junction with the A6. Although there are no signs to show this, it is recorded as such in official Department for Transport documentation.

The motorway passes right next to Manchester Metropolitan University for part of its route, whose buildings dominate the skyline, especially the shiny Business School building. (See photo opposite)

In August 2015, a 40 feet deep sink hole opened up at the eastern end of Mancunian Way in the non-motorway section where it becomes the A635. The repairs took ten months to complete and cost £6 million to put right, as a new sewer had to be put into place under the road.

It was the first elevated main road to built outside London and is 3,232 ft (985m) long. It was designed by G Maunsell & Partners, who also designed the Hammersmith Flyover in West London.

The A57 (M), Mancunian Way looking east. The whole of this motorway is on a flyover, which goes over all the major roads going into Manchester city centre from the south. Note the lack of a hard shoulder.

20 - THE A167 (M) – THE NEWCASTLE CENTRAL MOTORWAY

This is an urban motorway that runs north to south along the eastern edge of Newcastle city centre. It runs for just over a mile from the Tyne Bridge to Jesmond in the north and has several unusual features.

Almost as soon as it starts, it goes underground into a short tunnel under the Metro Radio building. Then further along it becomes a two tier motorway, with the northbound lanes going directly over the southbound lanes on a flyover. Plus, some of the exits actually veer off to the right, as opposed to the left as is usual motorway practice.

It was originally called the "Central Motorway East" as planners had dreams of building three motorways in Newcastle – one to the east of the city centre, one to the west, and one right through the middle of Newcastle! When it was first opened in 1975 it was named the A1 (M), then it changed into the A6127 (M) when the A1 (M) was moved further east through the Tyne Tunnel. Finally it was designated the A167 (M) in 1990 to fit in with the fact that Newcastle was now in the Motorway Zone 1 numbering system and not in Zone 6.

BRITISH MOTORWAYS – AN INTRODUCTION

Now known as the "Newcastle Central Motorway", it is one of the shortest motorways in Britain at just over a mile in length. It has a 50 mph speed limit and is a dual carriageway throughout its entire length from the Tyne Bridge to its end at the A1/A696 Kenton Bar Junction.

The A167 (M) in Newcastle-Upon-Tyne looking south, with the Tyne Bridge in the background. Note the short junction to and from the A186 on the left and the lack of hard shoulders.

21 - FORMER MOTORWAYS

With over sixty years of motorway construction in Britain it may surprise you that some motorways have disappeared from Britain's roads. It doesn't mean that they have been closed down for good, just that they have been downgraded from a motorway to an A road. Here are a couple of them.

The M10 was originally a spur off the southern end of the M1 when it opened in 1959, leading onto the A6 just south of St Albans. In 2009 there were several alterations made to the M1 south of Luton, including junctions 7 and 8. As a result the M10 was downgraded from being a motorway and instead became part of the A414.

Further south in West London, the M41 went for half a mile from Shepherd's Bush to join the A40 (M) Westway, as part of the "London Motorway Box". It was scheduled to continue to Willesden to the north and Battersea to the south, but was never extended. It became the A3220 in 2000 when Transport for London took over the running of this and several other London motorways. As Transport for London didn't have the legal power to maintain motorways, they were downgraded.

Other former motorways which have been downgraded to A road status include:-
The A18(M), a spur running from the M18.
The A41(M) Tring by-pass, now part of the A41.
The A102(M) in East London, now the A12 and A102, which lead to and from the Blackwall Tunnel.
The A6144(M) in Greater Manchester, now part of the A6144.

This was once the start of the M41 until 2000, when Transport for London took over the running of this and several other motorways in inner London. It was renumbered the A3220 and was originally part of the Ringways motorway scheme, which was eventually scrapped - apart from the M25.

LIST OF MOTORWAYS IN BRITAIN

Below is a complete list of motorways to be found in Britain, with the nearest town or motorway they start from or terminate at.

M DESIGNATED MOTORWAYS

M1 – London to Leeds/A1 (M)
M2 – Strood to Faversham
M3 – London to Southampton
M4 – London to Swansea
M5 – Birmingham to Exeter
M6 – M1 Rugby to Carlisle
M6 TOLL- Birmingham by-pass
M8 – Glasgow west to Edinburgh
M9 – Edinburgh to Dunblane
M11 – London to Cambridge
M18 – M1 to M62
M20 – M25 to Folkestone
M23 – M25 to Crawley
M25 – London orbital
M26 – M25 to M20
M27 – Southampton to Portsmouth
M32 – M4 to Bristol
M40 – London to Birmingham via Banbury
M42 – M5 to A42 Measham
M45 – M6 to A45 Rugby
M48 – M4 to M4 via old Severn Bridge
M49 – M4 to M5
M50 – M5 to Ross-on-Wye
M53 – Wallasey to Chester
M54 – M6 to Shrewsbury

BRITISH MOTORWAYS – AN INTRODUCTION

M55 – M6 to Blackpool
M56 – M60 to Ellesmere Port
M57 – Liverpool by-pass
M58 – M6/Wigan to Liverpool
M60 – Manchester orbital
M61 – M60 Manchester to M6 Preston
M62 – Liverpool to Hull
M65 – Preston to Colne
M66 – M60 to Ramsbottom
M67 – M60 to A628 Hattersley
M69 – M6/Coventry to M1/Leicester
M73 – M80 to M74 Glasgow by-pass
M74 – Glasgow to A74 (M) Abington
M77 – Glasgow to Kilmarnock
M80 – Glasgow to Stirling
M90 – Dunfermline to Perth
M180 – M18 to Grimsby
M181 – M180 to Scunthorpe
M271 – M27 to Southampton
M275 – M27 to Portsmouth
M602 – M62 to Salford
M606 – M62 to Bradford
M621 – M62 to M62 via Leeds
M876 – M80 to Kincardine Bridge
M898 – M8 to Erskine Bridge

A (M) DESIGNATED MOTORWAYS

A1 (M) – M25 to Gateshead (in four parts)
A3 (M) – A3 to A27 north of Portsmouth
A8 (M) – Bargeddie to Ballieston
A14 (M) – A1 (M) to A14 Alconbury
A38 (M) – Birmingham to M6
A48 (M) – M4 to Cardiff
A57 (M) – Manchester south by-pass
A58 (M) – Leeds inner ring road west
A64 (M) – Leeds inner ring road east
A66 (M) – A1 (M) to Darlington
A74 (M) – M6 to M74
A167 (M) – Newcastle by-pass east
A194 (M) – A1 (M) to South Shields
A308 (M) – M4 to Maidenhead
A329 (M) – Bracknell to Reading
A404 (M) – M4 to A404 Maidenhead by-pass
A601 (M) – M6 to Carnforth/A6
A627 (M) – Rochdale to Oldham
A635 (M) – Part of the A57 (M) Mancunian Way
A823 (M) – M90 to Dunfermline

One of the lesser known motorways in Britain. This is the M181, which leads from the M180 into Scunthorpe in North Lincolnshire.

Motorway signs for two short A prefixed motorways that start at junction 8/9 of the M4. They are both in Maidenhead in Berkshire. Note: The A404 (M) used to be the M4 before it was extended further west from Junction 8/9. A confusing junction to say the least!

WEBSITE LINKS

Disclaimer. Please note that these website links have been included in good faith. Neither the author, nor Hadleigh Books can be held responsible for the content and functionality of these websites.

Motorway Services – A website that catalogues and reviews all motorway service stations in the UK. www.motorwayservicesonline.co.uk

Pathetic Motorways – This website looks at the various motorways past and present that have unusual or strange features that the planners didn't foresee before they were built. It also looks at lost, former and renumbered motorways. www.pathetic.org.uk

Roads.Org.Uk - (Formerly known as "CRBD – Chris's British Road Directory"). This comprehensive website has details of all aspects of the road network of Great Britain, including its motorways. www.roads.org.uk

Sabre – The Society for All British and Irish Road Enthusiasts. A group of enthusiasts and professionals who discuss roads and share information including that of motorways.
www.sabre-roads.org.uk

The Glasgow Motorway Archive – Contains information on the history and development of the motorways in the Glasgow area.
www.glasgowmotorwayarchive.org

The Motorway Archive – This site has a vast collection of information and artifacts relating to the development of the UK's motorway system.
https://www.ciht.org.uk/ukma/motorways-by-region/

OTHER ROAD TRANSPORT BOOKS BY MARK CHATTERTON

This book is one of three introductory books on different aspects of Britain's Road Transport system.

The other two are:-

British Tunnels – an introduction by Mark Chatterton

It is available in the following e-book formats:
Amazon Kindle version ISBN 978-1-910811-58-0
E-Pub version for Apple I-Pad/Tablet or Kobi ISBN 978-1-910811-59-7
PDF file for a PC ISBN 978-1-910811-60-3

British Road Bridges – an introduction by Mark Chatterton

It is available in the following e-book formats:
Amazon Kindle version ISBN 978-1-910811-62-7
E-Pub version for Apple I-Pad/Tablet or Kobi ISBN 978-1-910811-63-4
PDF file for a PC ISBN 978-1-910811-64-1

BRITISH MOTORWAYS – AN INTRODUCTION

Due for future publication is:

Britain's Motorways by Mark Chatterton

This is a more detailed book than this one, covering all of Britain's motorways, from the M9 and M90 in Scotland down to the M20 in Kent and the M5 in Devon. It contains detailed entries of almost seventy motorways, as well as over a hundred photographs with detailed information on each motorway. There are many other facts and figures about Britain's motorways included, such as motorways in films and on TV, Smart motorways, events which closed motorways and motorway service stations.

It will available as both a printed book and an e-book

ABOUT THE AUTHOR

Mark Chatterton has travelled the length and breadth of Great Britain on every motorway that exists in Britain today. On his travels, he has amassed many photographs and much information about them. This is his first book on motorways, which is a precursor for a more detailed and lengthy book on Britain's motorways. By the way, his favourite motorway to drive on is the M45; or maybe the M25……. at around midnight!

Website - markchatterton.com

Facebook Page – Mark Chatterton - Author

© Mark Chatterton 2016/2021

Hadleigh Books, Church Road, Hadleigh Essex SS7 2HA UK

www.hadleighbooks.co.uk

Printed in Great Britain
by Amazon